I0803882

HOW TO $AVE ON

HOME DRIVEWAY

INSTALLATION OR REPAIR

Alfred Earl Edwards, III

FAMILY ORIGINALS
PUBLISHERS

Family Originals Concepts Corporation
Beachwood, Ohio

How to $ave on Home Driveway Installation or Repair

Designed and illustrated by Myrick Holliday Associates
Edited and produced by A&W Designer Typography

Published by Family Originals Publishers, Inc., Box 250, 13944 Cedar Road, University Heights, Ohio 44118.

FOREWORD

Let me help you save money on your driveway, now and for the years to come!

I wrote this book to save you money in an area for which there is little or no information. Every year, homeowners spend millions of dollars to replace or repair their driveways. With this kind of money involved, we should become educated and avoid being ripped off.

We need to learn more about contractors, get to know what they know. Driveway repair or replacement is a very expensive project for a homeowner. If done correctly, it can beautify your property considerably. The information you are about to read will open your eyes to controlling cost, understanding your contractor, understanding how various materials are used, You will be able to supervise installment and do your own maintenance and repairs.

I have made mistakes over the years, but I also LEARNED a lot by listening to my customers, doing the necessary research, going to the seminars (the few given). This book was put together so YOU may learn from my mistakes, research, and experience.

So, sit back and enjoy the ride to your new driveway, your new "PROPERLY INSTALLED" driveway. When you have finished, think about how you will spend the money you save.

TABLE OF

CONTENTS

INTRODUCTION

This book is intended to help the consumer become an expert in choosing the right professional assistance for home driveway and outdoor paving. It also includes a step-by-step guide that illustrates the procedures needed to ensure a well-built, long-lasting, and attractive paved driveway. This pictorial guide will be of value for both professional paving contractors and do-it-yourselfers and will provide homeowners with a means of judging the quality of a contractor's work as it is going on.

My father installed driveways for a living. He also built homes the largest part of his work was installing driveways. A big part of the conversation that went on in our home was about "driveways." How to install them? How to price them? What parts should be removed before resurfacing a driveway? How to build a solid base? What materials to use? How to make sure that the drive would hold up, etc.

To me, it was like being in school. By the time I was 13 years old and in junior high school, I was working with my Dad every day during the summer. My friends would play and I would work. Even when I was very young, I enjoyed working on homes and especially enjoyed watching my father enjoy his work. He showed customers how to save money and showed me how to be honest. He was very honest in his work; some in the business would say too honest. When a contractor installs the right amount of material in a driveway, the other contractors called it too honest. He would not cheat! He really never became rich and I grew to love him for being that type of person. He taught all six of his boys how to work with our hands as well as our minds. How to do the best we could while working. *He drove us very hard!*

So I wrote this book about my experience as a home-driveway contractor: how to avoid mistakes and what I can do to save people money on their home driveways-one of our biggest home investments.

In 1971, my father moved my mother, four younger brothers, my sister, and me to a small town outside of Cleveland. I learned a lot in those years. At that time, most people there did not have paved driveways. My father and his partner thought they could pave the entire town and surrounding townships. We certainly tried!

The people there were pleasant to work for, and we did not mind doing extra free things. They would prepare lunch as well as refreshments. Watching our company work was an education for those customers. My father, his partner, and their ten boys, a black and white family united, worked together as one. Many were shocked at first, then they relaxed and came to trust my father's and his partner's judgments, opinions, and the quality of their work. Their reputation grew and their company, *Edwards and Fowler Construction and Paving*, also grew. They built several homes and installed 200 driveways or more each year through 1975. Without doubt those were the most important years of my life as far as construction is concerned. Learning to work with and please the public is not easy to learn.

I have now spent 24 years in construction. Installing driveways is one of the most challenging areas of construction I know. I will never get rich installing asphalt, but I enjoy it. I don't do a lot of the things I saw larger companies that I worked for do. As an independent contractor, I've never gotten involved in the politicking needed to get the big contracts. I stayed home and learned more about driveways than most contractors in the business. I love seeing an old beat-up driveway being brought back to life again and the smile on a customer's face when he sees a new driveway.

A driveway is usually the last step in finishing a new home. And you never even notice it--unless it's not there or in bad condition. When you drive in and get out of your car you look at the house or the paint job or the landscape. Then, one day you go out to get in your car or cut the grass, you look down and see a big crack running across the driveway and you say, "man, oh man, what's wrong." Looking further you notice a small hole developing or an allegation area (a network of cracks). Now, you have a problem. This could be one of the most expensive repairs that you will have with your home, outside of a new furnace, a new roof, or basement waterproofing. Every homeowner will repair or install a new driveway at some time.

In the case of a new home, home contractors don't give a lot of attention to the driveway. They are too busy looking for the $50,000 or more they make building the home and try to nickel and dime the asphalt or cement contractor down to the lowest point possible. Believe me or any driveway installer when we tell you that. The building contractors will say, "Cut back, just make it look good," or, "Make it pass inspection." I don't like that--period. I have owned 3 homes in my time and, being a homeowner, I would not want that done to me.

1

THE NEED FOR INFORMATION & HOW TO SAVE MONEY

NOTES

1

THE NEED FOR INFORMATION & HOW TO SAVE MONEY

Knowing that a driveway is one of the five most expensive home-repair improvements, I began to listen to many questions from my customers about their driveways. I recognized the need for information for the general public on the subject. Contractors often sell a driveway installation or repair without properly explaining to customers why they should or should not install an asphalt or concrete driveway for their homes.

Over the years people have asked me how much they should invest in installing or repairing a paved driveway. What are the chances of that driveway holding up? If it does not hold up, what are the customer's options? What will they save by having a new drive installed? What is lost by neglecting repairs?

Starting with these questions, I began to gather information on what to do to install, replace, or repair your driveway, and how to save money doing it. I also began to study what to do and look for in choosing a good contractor, if that becomes necessary.

In the times we are faced with today, it is important to find ways to cut every expense possible! One of our most important expenses, aside from our families, is our home. I have seen many publications on home improvements and how to buy and sell homes. As a driveway and parking lot contractor, I'm tired of seeing this issue constantly ignored by authors of such fine and detailed home improvement books. These books do make very good points about how to buy and fix up or decorate your home, and they are needed. But, maintaining a part of surroundings that make a home stand out is also very important. Everything else being equal, which would *you* be more likely to pay more for: a house with or

NOTES

without a paved driveway? After 22 years of experience and a great deal of research, I know that at least 90 percent of all homes have driveways; some of which are built especially to allow a car to pull up close to an entrance.

I have built driveways from the sub-base up and know how to put material into its proper perspective. Remember, in construction, the word "specs" is often mentioned along with the blueprints. When it comes to construction, materials must be specified when the blueprints are drawn.

At times it seemed that my father would go overboard on materials when building a driveway. However, when he would drive me by different homes and say he worked on that driveway 18 or 20 years ago, I would say to myself, "That long ago! They must have done it the right way." Today I see many driveways falling apart after only 10 years. Something must be wrong. That is why I am trying to explain to you, the public, that there are hundreds of millions of dollars being spent each year on asphalt alone. People can save money by following just a few steps that I'll point out as we go along.

This is the first book that I have ever heard of that can show a person "how to install a driveway or a parking area" or any of the following: How to save money. What are the materials made of? Why does a driveway not hold up? What to look for in a contractor? How to do preventive maintenance. Why people shouldn't blame the contractor in MOST cases, etc. I want to let the public know that this driveway business goes unnoticed each year; yet each year we spend millions of dollars, hundred of millions, billions over the years.

There is no solid information in the bookstores to help the public. Most people with a driveway problem have no idea how to pick a contractor. When my company was just getting started, people would tell me, "I have a problem with my driveway and picked your name out of the *Yellow Pages* or newspaper ad, I need a price on my driveway." Most customers never thought to ask questions such as: "Why did my drive fall apart?" "How can you save me money?" What can I do in the future to save my driveway?"

NOTES

They say, "time is money." **I am talking about $1,000 or more in most cases!** If I can save you $500 or more, would you be happy? What about a driveway that cost $1,500 or $3,000? The most expensive driveway I ever put in cost $7,000. I remember a driveway my father and his partner installed that cost over $10,000 in 1972. And this was for asphalt; the price would have been even higher for concrete. So you see that a driveway can get up there in cost.

You might think that the kind of person who spends a $1,000 or more for a driveway can afford to do things right. This is not always the case. In the city where I live and work, city inspectors can issue orders to install new driveways or bring others up to code. If the work is not done within a certain time and according to the city's standards, the homeowner can be called into court, fined, and even jailed. I've seen people who have had to spend their tax refunds, vacation money, or their children's college education money just to bring their driveways up to city code requirements. This may not be the situation everywhere, but I have seen many communities where it does happen.

The fact is that most people want to protect their home investment. That's why I took the time to write this book in the hope that it will help you, a family member, a friend, a neighbor, or a senior citizen. I don't want you to get ripped off or to spend money in an area of your home where you might not have to. However, if you do find that you need a new driveway or resurfacing job or repair work, look at all the chapters, read carefully, and follow the steps that show you how to cut back and save some of your hard-earned money.

2

FOUNDATION

How to Install and Build Your Base Properly, What Materials to Use, How Much to Use, and Why It Is Important

FIGURE 1

FIGURE 2

FIGURE 3

FIGURE 4

FIGURE 5

2

FOUNDATION

A well-made asphalt driveway, like anything else we build, needs a good foundation. If you are putting in a new driveway, insist that the contractor first installs a solid sub-base.

First, dig a shallow bed for the driveway, 12 inches deep. Then lay the sub-base, consisting of oversized limestone rocks, 3-4 inches or more in diameter (grade B19 or 304) (figure 1). Whether you do the work yourself or have contracted it, you need to know what to order. Check with the stone docks in your area and make sure that you get the right stone for your driveway or parking area.

Limestone is almost as hard as pavement, but it will shift unless it is paved over. At least 6 inches of limestone should be spread to form the sub-base. You can use a rake to level the limestone after it has been spread by a tractor. If you do the work yourself and by hand, have the delivery truck dump it in small piles, 2 to 3 feet high. It is important to compact this limestone layer. (See figures 1-5 and the Step-by-Step Guide, "Sub-Base," at the end of this book; referred to as the "SBS Guide," hereafter.) You can get the tools and equipment you need from a tool rental agency or a hardware store.

The stone base is the strength of your driveway. It is your foundation, the most important part. Remember the old saying, "A house built on a weak foundation will not stand." The same is true for a driveway. If you remove the mud, grass, and weeds and install the base yourself, you save money! You are also making sure you have the best base possible. The cheapest place to buy your stone is from a stone dock (when it first comes in) or from a quarry yard. A builder-supply store charges double or even three times the cost at a stone dock or quarry, which will generally charge $3-$6 a

FIGURE 6

FIGURE 7

FIGURE 8

FIGURE 9

ton. Since you may need 10 tons or more of stone, you can save $100 or more by buying it directly.

Calculating the Materials You Need

First, figure the square footage of your driveway. If it is 10 feet wide and 100 feet long, you have 1,000 square feet (multiply the width by the length to get the square feet). Call your supplier and give him your square footage, he will be able to put your order in for the exact amount you need.

One ton of limestone slag, 2 inches thick, should cover 80 square feet. If you are looking for 6 inches of thickness, remember, one ton covers 80 square feet at 2 inches thick. That same calculation is used when we move onto the asphalt. The supplier should be able to help you once you've taken your measurements.

Once you or your contractor levels your stone base with a rake, make sure it is compacted with a heavy roller. You have to compress the stone until it is solid all over, with no stones sticking up, if possible. If you install the sub-base yourself, I recommend a 4-6 ton roller or heavier because the stones are large and you need the weight. If a contractor does this work, make sure the sub-base is properly installed. It must be solidly compacted to support the asphalt.

The Finish Base

The next stage in laying a good foundation requires putting down a finish base over the sub-base. The finish base uses smaller rocks (2-3 inches in diameter) laid directly over the sub-base. The same procedures are followed: spread the stone evenly and compact them with a power roller. (See figures 6-9 and the SBS Guide, "Sub-Base."). Figure 10 gives a detailed cross-section showing the sub-base, finish base, and asphalt layers.

Repairing or Replacing a Driveway

If an asphalt driveway has been resurfaced three times or is

NOTES

badly deteriorated, you probably need to have it torn up and removed. Hire someone with their own tractor to rip it out and haul it away. A driveway of 1,500 to 2,000 square feet should cost no more than $300-$400 to remove.

It would be a good idea before arranging for the removal of the old pavement to dig up a small section or two to see how much of a stone base you have already. Watch for mud in soft or weak spots--it will have to be removed and replaced with limestone. If the base is too thin or missing in spots, you can arrange to have the new material you need ready at the same time the tractor is being used to remove the old driveway. That way the tractor operator can help you spread and compact the base right away.

If a contractor is repairing your driveway, check the base at the same time he does. If there are soft or muddy spots, insist that they be filled in with a good base. Don't take anyone else's word that the base is in good condition, double check for yourself. Once the job is done and paid for, it may look good; you may have a guarantee for a year; but who will pay if trouble develops two or three years later?

The point that I'm trying to make, if you haven't gotten it by now, is that a solid stone base is absolutely essential for a driveway, whether it is covered with asphalt or concrete.

A Scotsman, John Loudon McAdam, who lived in the last century, is famous as a road builder and as the inventor of the material named after him, "macadam." His roads were constructed of crushed rock or stone packed into thin layers. He was also the first person to recognize that dry soil supports the weight of traffic and that pavement is chiefly useful for forming a smooth, drivable surface and for keeping the soil dry.

My years of experience have proven to me that McAdam was right. Crushed rock packed into thin layers forms a good base and a smooth pavement covering keeps the base dry and in place. I repeat, a good base installed correctly on solid, prepared ground is the "LIFE" of your driveway.

One of the interesting things I found while doing

NOTES

research for this book is that the ancient Romans may have been the best road builders of all time. Roads that the Romans built back when they were the major power in Europe still exist today.

Not too long ago, I read in my local newspaper that the federal highway administration was going to spend over $100 million and the local state, city, and county governments were investing another $42 million on highways in the three-county area where I live. $142 million dollars! If the Roman roads could last over 2,000 years, something is very wrong with modern technology and industrial building. The problem comes from the way the base is built for these highways and from the materials used on their surfaces.

3

PIPES & DRAINAGE

NOTES

3

PIPES & DRAINAGE

The Secret of Long-Lasting Paving, Planning Run-Off and Drainage Areas

The Romans were also the first people to build roads with planned drainage. You must have good drainage for your driveway! I have noticed in installing driveways that poor drainage can cause a lot of problems for a driveway.

My dad taught me the importance of good drainage. I remember working with him on a home construction project near Huntsburg, Ohio. As he was getting ready to install a septic tank, he discovered that the lot where the house was being built had poor drainage and that it was very wet in the backyard. He insisted that the homeowner have us lay PVC pipe all around the property to drain the water as far away from the house as possible. His plan worked out fine, and the backyard became a good area in which to walk, run, and play.

It was a lot of work, but I learned something from that experience. We started using the same idea when we installed driveways, to take the water away from the driveway, especially from the edges. At least 75% of the time, the first area to crack in an asphalt driveway is along the edge. By making sure the edges stay as dry as possible, moisture will be kept out and the driveway will last. The edge is the weakest point of a driveway; you have to protect it. Why spend $1,000 to $3,000 of your hard-earned money and then waste it by taking a shortcut that will save you no more than $100 or $200 to protect the investment. Good drainage can add another 5 years or longer to the life of your driveway; that averages out to a little more than $1 a month.

Water can exert a lot of force and do a lot of damage

NOTES

to a construction project. We waterproof our basements to protect our houses and that costs much more than providing good drainage for a driveway.

When you or a contractor install the base, remember that the Romans made sure the water would drain off and away from their roads. The same applies when it comes to your driveway:

(A) Don't allow the run-off from your house to drain onto the driveway.
(B) Be sure that the drainage from the driveway is carried far enough away so it does not run back under or next to your asphalt, concrete, or brick driveway.

Keep it dry! Remember, the dry soil underneath supports the weight of the traffic on it. A wet base or surface will not hold up. Lay PVC flexible pipe with holes in it along the side of your driveway on both sides to help absorb the water. Dig a trench 10 inches deep, put a 4 inch PVC drain pipe with bank gravel under the pipe; then cover the pipe with the same soil you dug out. Or, you could use a solid piece of PVC with a drain crock if the driveway is sloped to allow the water to flow to the drain. If you decide on this arrangement, keep in mind that some water will run off the drive and the soil near the edges will get wet. A drain running under the length of your driveway is the best way to protect your driveway edges from cracking and eventually breaking apart. It will help save your driveway.

In many rural and some suburban areas, driveways have what is called a cover pipe at the end of the driveway. It allows the water in ditches to flow under the driveway. Often ice will form under and in the cover pipe during the winter and it will lift up because it is hollow; then a problem starts. The blacktop or concrete above lifts, breaks, or cracks.

It helps a lot to lay a 4 inch PVC drainage pipe on each side of the cover pipe. Bank or beach gravel packed

NOTES

around the cover pipe is also helpful. You can now get PVC cover pipe 12 to 14 inches in diameter; it is better than the old metal pipes.

Water standing along the drive will eventually soak under it and will kill the driveway. Please keep that in mind. Install a drain system along the edges if your driveway was installed in the last five years. It will not hurt and may add years of useful life. If you are installing or resurfacing your driveway, make sure that water will not pool under or beside it and that the edges are protected with a drainage system of some sort!

4

ASPHALT DRIVEWAYS

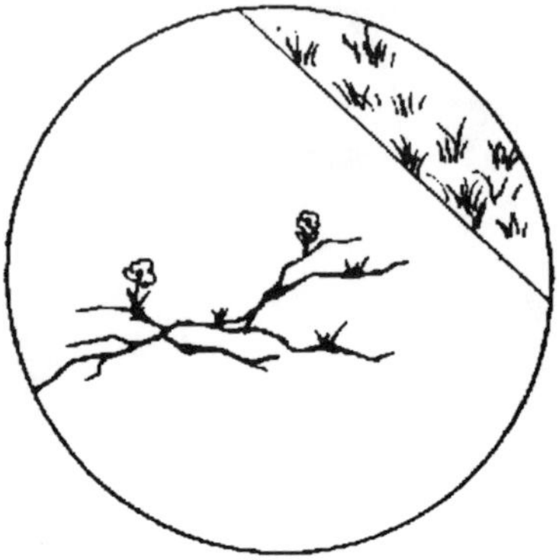

FIGURE 11

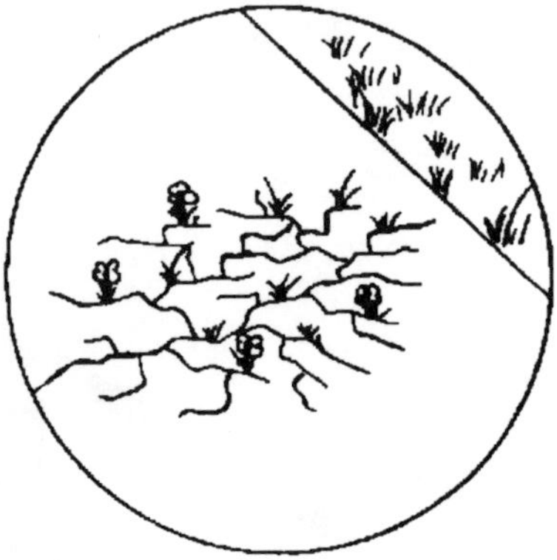

FIGURE 12

FIGURE 13

FIGURE 14

FIGURE 15

4

ASPHALT DRIVEWAYS

Basic Principles of Asphalt Driveways, Weeds and Mud

> ***Asphalt:*** a brown to black bituminous substance that is found in natural beds and is also obtained as a residue in petroleum refining and that consists chiefly of hydrocarbons; an asphaltic composition used for pavements and as a waterproof cement.--*Webster's Ninth Collegiate Dictionary* (Merriam-Webster, 1987).

Asphalt is used today to surface most of our highways, most parking lots and over half of new driveways. It is a by-product from refining crude petroleum that is mixed with crushed stone, chemicals, sand, and tar, then heated and installed while hot. Asphalt is not going to last forever, no matter what a contractor might say. There are, however, some things that can be done to make it last longer. The first thing is to make sure that all the mud has been removed from the area where it is being installed. *All mud with no exceptions!* Asphalt is not as strong or as rigid as concrete. Any soft spot under it will allow it to sag. Carefully go over the area where it is to be laid and don't let a contractor tell you it's okay. Asphalt laid over mud or soft soil will sag and crack. Forget about jackhammers, you could take a pick and dig straight through and lift pieces of it with your hands. If an asphalt driveway is installed over mud, you can feel it move almost like a wave or as if you have a blob underneath your feet when you walk over it. The area will not get hard!

NOTES

Dandelions: The Driveway Killers

Never install an asphalt driveway or parking area over grass, and especially not over a DANDELION PLANT. Dandelions have deeply notched leaves with bright yellow flowers on a hollow stalk. They bloom in the spring, and you will see them if you have them in your driveway or parking area. Make sure that you or the contractor digs them out from the ROOT UP; do not just cut off the top of the plant. (See figures 11-15 and the SBS Guide at the end of this book.) They will grow right through the asphalt with no problem at all. They are strong, very strong. When they are removed, the area should be sprayed with a weed or vegetation killer. If you object to chemical herbicides, spread rock salt or table salt where they grow. Don't think that the asphalt will kill grass or dandelions for you. It won't work! If you've ever seen an abandoned house that no one is keeping up, you can see grass and weeds growing through cement or asphalt pavement. Well, in a new or resurfaced driveway, weeds and grass will grow through your new surface as well. It is very important that you take out dandelion plants from the root up and don't miss any. Remember, dandelions are as damaging as mud for your driveway. They will also separate the layers of black top which holds up better when it is one continuous layer.

5

INSTALLATION OF ASPHALT

NOTES

5

INSTALLATION OF ASPHALT

Installation of a new asphalt driveway . . . properly!

One thing I remember: after we moved from the city, my dad thought of using two layers of asphalt for a driveway. His reason was simple: it added **strength**. Have you ever noticed that when a street is being repaved, the workers put down one layer, compress it with a roller, then come back and put down a second layer of material? Or a paving project may not be finished and sewer, meter, telephone, or manhole covers are too high. When you hit one of these covers with your car, you swear and think the construction company did a lousy job. Here you are, driving on brand-new black top, and it's worse than hitting potholes. The explanation is simple: you are driving on what is called "binder asphalt," the first layer to go down over the stone base. In the region where I work, it is called #402 or #301 binder asphalt and is also used as the first layer for adding strength to a driveway. This binder layer adds durability and meets the standards (or spec's) of the department of transportation.

While working for a company that installed streets and highways, I learned more about installing two layers of asphalt. It was about this time (1982) that I first worked with rubberized asphalt, which is asphalt with rubber in it. I was a *raker*, the guy who works behind the spreader box who ties in the joints and helps keep it level for the next path. The raker is like a concrete finisher or concrete puller. It takes years and a good eye to learn how to do this properly, and I am forever grateful to my dad for first teaching me. It took five years for me to become good at this, but now I can rake a driveway just as level by hand as a spreader box could. I like to make it neat and pretty for my customers. It's like painting a picture; I love this particular

NOTES

job now just as much as when I started in the business.

Twenty-eight years ago when I first started working for my dad, I wanted to be an asphalt raker and he said, "No, you're going to start working on the grade crew and learn how to prepare a driveway for installation." I didn't like that job at all. Nevertheless, I was only thirteen years of age and I was earning $60-$80 a week so I did it and began to develop an interest in learning how to grade the correct way. You can learn also and get enjoyment out of working with your hands. It's also a good way to keep in shape.

I want to finish talking about binders in asphalt driveways. As I described above, the binder forms a connecting layer between the stone base and the top layer of asphalt and prevents water from rising up under the top layer of asphalt. It also supports and strengthens the top asphalt layer and keeps it from separating from the stone base.

Installing the Binder

The binder is asphalt with larger stones in it. It should be spread 2-3 inches thick over the stone base (never less than 2 inches thick for a driveway). After it is down, it needs to be compressed with a heavy roller. Then the driveway is ready to have the final, finishing layer of asphalt installed.

A lot of contractors want to hurry when they do your driveway so they can move onto the next job. Why rush a $1,000-$3,000 investment? The contractors can install a thousand-square-foot driveway in 3-1/2 hours if they skip the binder layer. Don't let that happen! If you let them get away with it, you may as well take your money and throw it in the garbage. A 2-3 inch binder will help that driveway last many years to come. I have seen one-layer driveways fall apart within three to five years or less.

Once again, your new asphalt driveway should have the mud removed, a drain installed, a good stone base, then a layer of asphalt as a binder. Like they say "don't

NOTES

leave home without it"--don't install without it!

A builder or home improvement person doesn't often give a lot of attention to the driveway. They build you a $100,000 home and give you a $200 driveway. Within five years you'll have to spend $1,000 to $1,500 or more to repair your driveway. If you have to take out the old asphalt, the job will cost $2,000 or more. So, which is fair? The contractor might have built you a great house, but does it make sense if you have a driveway that looks like it is ten to fifteen years old within three to five years. If your top layer has holes in it or falls apart, the base underneath gets wet, mud spreads, the drive will begin to weaken, sag, and crack. Then you have to spend money on a new driveway instead of a vacation or new car. Maybe you were just ripped off because no one told you what to watch for; *well, I am!* Roads are built to take a pounding, and your drive is like a road. You pay for both, one way or another. If a binder layer is good for the streets, it should work even better for your driveway.

I used to hate raking the binder layer and then having to do the same thing for the finish layer. It seemed like we were doing the job twice and losing money on the extra work; but that was not so. My dad and his partner did not advertise; all their jobs came from referrals because of the quality of their work. A lot of their customers lived outside of cities and delivery of fuel oil and other things would come in large trucks. The two-layer driveways that my dad installed held up to the loads while single-layer drives built by other contractors fell apart. Whether or not you have a heavy vehicle on your property, you still need two layers of asphalt in your driveway. When an old asphalt driveway is torn out and a new one put in, be sure to get two layers of asphalt: a binder and a top finishing layer.

Now, remember that street construction we were complaining about at the start of this chapter because the manhole covers were above the surface? Go back another

NOTES

day and everything is smooth; there are no more bumps. That's because the top layer has now been installed to match grade to the drains and has been compressed to a proper compression.

The Finish Layer

The finish layer is the one I enjoy installing the most. It looks attractive. I know the job is almost complete and that I will be getting paid and moving on.

The top layer is the fine finish layer with small stones to give it a nice smooth finish. Where I live, the material for the finish layer is called *topping* or #404. You will need one ton of asphalt for every 80 square feet to give a finish layer 2 inches thick. Demand that your contractor show his receipts from the asphalt plants! This way you can see how many tons of asphalt went down on your driveway to make sure you get your money's worth. A raker can stretch the asphalt and make it look thick by building up the sides of the driveway. You should take a ruler and stick it in the asphalt to measure its thickness while it is being raked and before it is compacted. Make a sample measurement every ten or twenty feet and especially at the end of the driveway where some contractors thin or stretch it. You have to protect your investment and make sure you are not getting taken. When you measure it, the top layer should be 2-1/2 inches thick so that when it is compacted you will have a top layer 2 inches thick. If you find that it is only 2 inches thick, it will compress to 1-1/4 inches after compacting and that's not enough asphalt.

Compacting

When installing the top layer of asphalt, make sure you get the correct amount of compression. Asphalt develops air bubbles in the mixing process and these form pockets once it has been installed unless it is compressed. A power

FIGURE 16

FIGURE 17

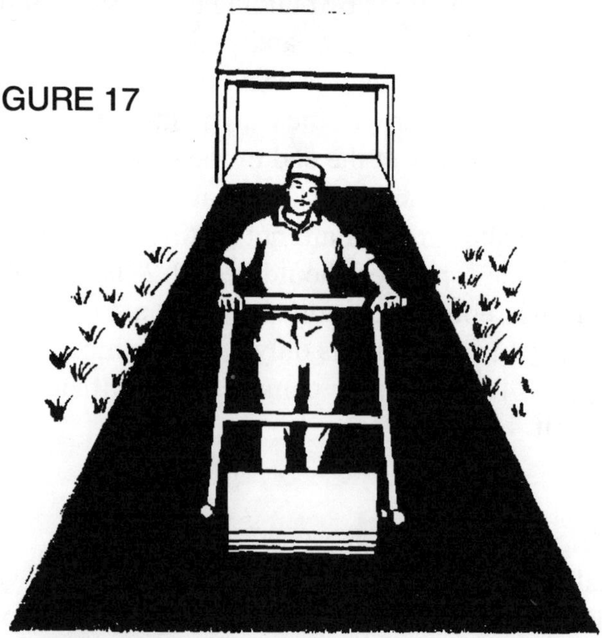

roller can compress the top layer and form a tighter bond with the binder layer. There is also a big difference between using a hand roller and a power roller. (See figures 16 and 17 and the SBS Guide, "Compacting.") You don't want a contractor to use a small hand roller only when he installs your driveway; don't let him talk you into it. The drive will not hold up--period! One of the most valuable lessons I learned at a seminar on asphalt installation was that if you don't compress the asphalt then don't look for the driveway to hold up. Water will get in between the asphalt layers and freeze in the winter. Water can rise from the bottom to the top; if you have the correct amount of compression with no air pockets, your driveway will be solid and will have a better chance of holding up.

Insist that your contractor use a power roller to compact the finish layer. A hand roller is not heavy enough for proper compression. The finish will not be the same; your drive will develop cracks and will not hold up as long if only a hand roller is used. There are times when a hand roller can be used: to start the first five feet of the driveway while the asphalt is still too hot for a power roller and to smooth out lines that the power roller will leave (see figure 18).

I know a contractor that is no longer in business in my area. He would use a hand roller, not a power roller. He got away with it for a while because the people he serviced did not know the difference. Customers would go to work and when they would come home, they would see their driveways all ready installed. The drives looked good for a while, but pretty soon people would notice their driveway unraveling and falling apart right in front of their eyes. In time the asphalt would peel all the way down to the slag. So be careful. You should make sure that the stone base, the binder layer, and the top layer are each compressed with as much weight as the driveway can take (see figure 19).

Since asphalt, unlike concrete, is not installed using

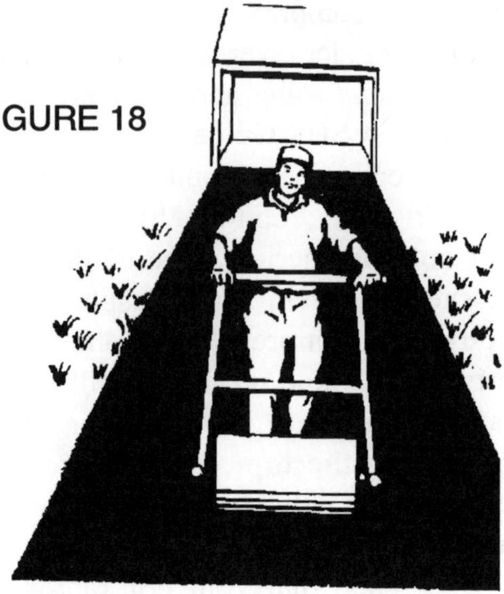

FIGURE 18

FIGURE 19

FIGURE 20

forms, the edges should be compressed with a hand tamp as well; they will hold up better. See that your contractor uses a heavy all-steel tamp for the edges. (See figure 20 and the SBS Guide, "Tamping.") When you repair potholes yourself, the lighter wood-handle, steel-base tamps that you can get at a hardware store will work perfectly fine.

I have had to dig into asphalt that was installed years ago and have found that it was still as shiny as coal; it was still wet and loose in the middle because it had not been properly compressed. Don't make the same mistake.

When you are having your asphalt installed, make sure the bottom and the top layers go down on the SAME DAY to allow the hot tar in the asphalt to bind the two layers together. If you wait and the contractor has to come back the next day, then make sure an emulsion-based tar adhesive or hot tar is used to bind the two layers together. This is very important.

6

ASPHALT RESURFACING

FIGURE 21

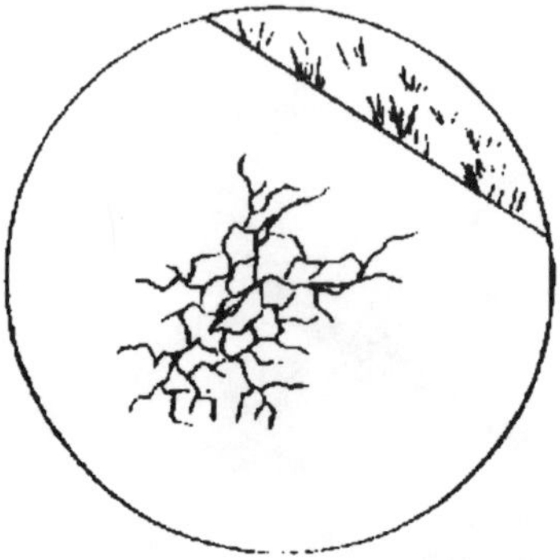

FIGURE 22

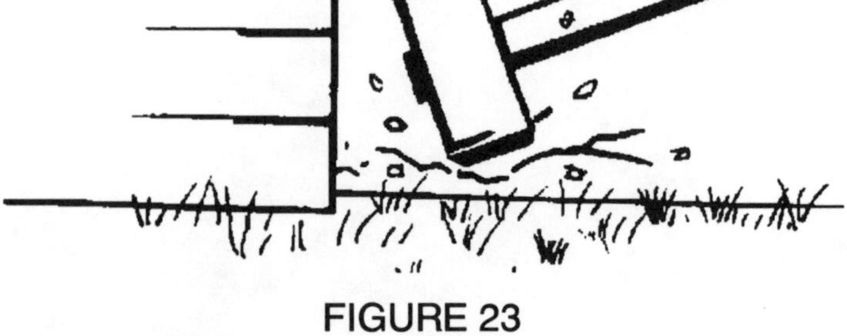

FIGURE 23

6

ASPHALT RESURFACING

Resurfacing of an asphalt driveway . . . when to resurface, how to do it, and when not to resurface.

One of the most important things when it comes to resurfacing, is to make sure that the asphalt that is currently down is able to support the new layer you are about to install. The new layer will not hold up if the original surface of the drive is 75% covered by allegation cracks (see figure 21 and the SBS Guide, "Asphalt Cobweb Crack".) If this is the case, do not resurface the driveway; dig it up and remove the old asphalt.

When you do resurface by covering one layer of asphalt with a new layer, the entire driveway should be clean and covered with hot tar or a cold emulsion-base tar adhesive (see figure 22). If not, the asphalt will not bind and will not hold up. As an experiment, put sand or dirt over an area of your driveway, then pour some asphalt or tar over that area to see if it sticks or binds to the old surface. It will not! The adhesiveness will not be there. Remember this experiment, whether you are doing the work yourself or have hired a contractor.

Resurfacing

Resurfacing a driveway should not cost over $1 a square foot, unless you have a driveway 500 square feet or less, which is not often the case. Follow these steps very closely and your driveway will hold up longer, with fewer repairs. When a contractor tells you that you need to resurface, be sure he follows these guidelines:

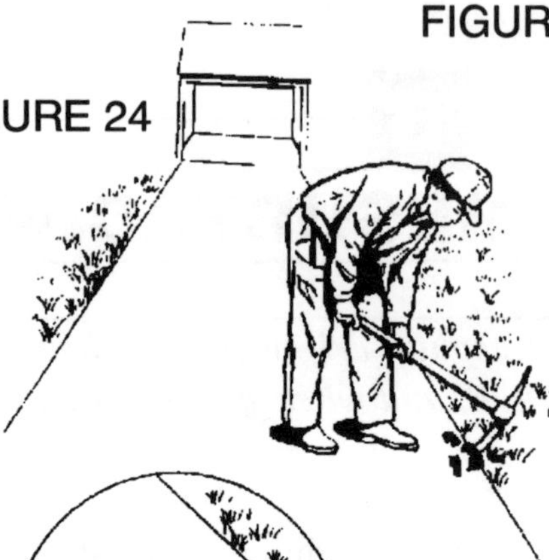

FIGURE 24

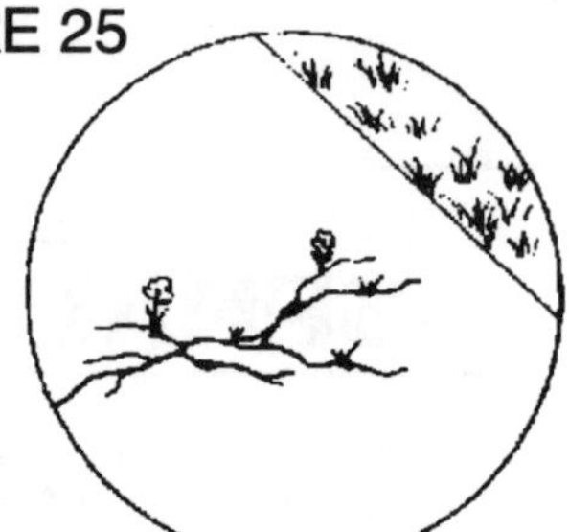

FIGURE 25

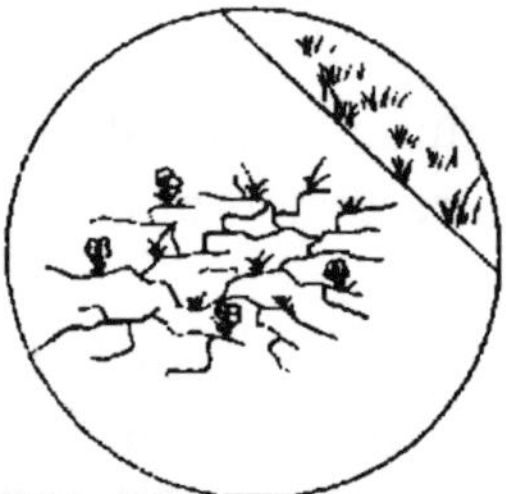

FIGURE 26

FIGURE 27

FIGURE 28

FIGURE 29

FIGURE 29B

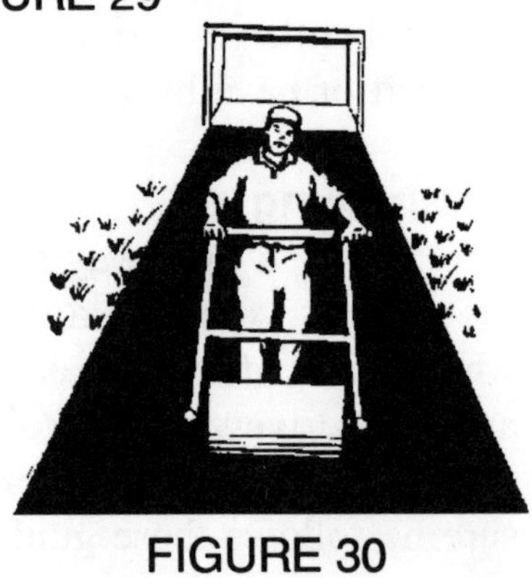

FIGURE 30

He should remove any broken area down to the bottom to see what is wrong (see figure 23 and the SBS Guide, "Leveling and Removing Asphalt".) If, for instance, out of 1,000 square feet only 100 square feet has mud underneath, only this section will have to have the mud removed and be rebuilt. You could even do that yourself. Make sure he trims the sod and grass back to find the edge of the original driveway (see figure 24).

Remove all crumbled and broken areas. They are weak and no longer serve as a base for the new asphalt. Cut them out completely, just like cancer, remove it all! As I mentioned, remove all grass and weeds and kill any roots or seeds. (See figures 25-28 and the SBS Guide, "Weed Removal.")

Build the new or removed area back up with a solid stone base and new asphalt. Compress the new section separately with a roller (figure 29-30) unless it is small enough to use a tamp (figure 31). Clean the driveway thoroughly, then apply an emulsion tar adhesive for binding the old asphalt to the new. Make sure the binder or tar is spread over the entire driveway.

Emulsion tar adhesive costs $2.00-$2.50 per gallon, and it is important. Some contractors try to use one 5-gallon bucket for the entire driveway. Don't let a contractor water down the emulsion, it will lose strength. It is important to cover the edges completely. (See figure 32 and the SBS Guide, "Emulsions, Tar Painting, and Seal Coating.") We called my brother Joe "the painter of emulsion." My other brothers Derrick and Reggie would get upset with him, feeling he was putting on too much emulsion. So did I. However, we were wrong! Joe went to

FIGURE 31

FIGURE 32

FIGURE 33

FIGURE 35

FIGURE 34

a seminar with me and we both learned how important it is to use enough tar or emulsion.

If you have cracks in your existing asphalt or concrete driveway, use a heavy-duty crack filler to fill them and close the surface before you put the emulsion down. After you have filled all allegation areas and all holes have been dug out and rebuilt, check out the drainage as well. Contractors sometimes overlook these small steps. So, read and follow along.

Now, once the emulsion is down, make sure your contractor uses *virgin asphalt*. I hate recycled asphalt. It's something that is new over the last 10 years or so. I feel it's wrong for driveways because it is something that was already used, re-cooked, then put back down. It stinks in my book! Virgin asphalt costs a little more, around $23 a ton for recycled as opposed to $27 per ton for virgin asphalt, but I would rather pay $270 for 10 tons of top asphalt from a virgin plant. I feel it will hold up longer.

I may be too tough on recycled asphalt; but in my opinion, you will be far better off in the long run by insisting on virgin asphalt. As a raker I am the last one to work the asphalt before it is rolled or compressed, and I can see the difference in driveways with virgin and with recycled asphalt.

Make sure that you get 2-1/2 inches of asphalt for your resurfaced drive. On a new driveway, there should be a total of at least 4 inches for the binder layer and finish layers combined. Before your contractor starts the compressing and tamping of the edges of the driveway, check the thickness of the installation. I repeat that warning because it is very important. Make sure that the edges are tamped with an iron tamp, not the raker's rake. The rake is also called a "loop" or "asphalt" rake. (See figure 33 and the SBS Guide, "Tamping.")

For the final step, be sure the contractor uses a hand roller only to take the lines out. Don't let the contractor finish the driveway without using a power roller

NOTES

weighing one ton or more. If he wants to finish with a hand roller because his power roller is not working, that is fine, provided the entire driveway has been compacted with a heavy power roller (figures 34-35). Put that requirement in your contract. Put all of this in your contract!

Summary

1. Excavate all broken areas, rebuild from the base up.
2. Remove all grass and weeds from the root up. Kill the areas with a vegetation killer and weed killer, in that order. A weed killer alone won't kill the grass.
3. Install a new stone base if needed in bad sections, then lay binder asphalt over the compacted stone.
4. Clean the drive completely.
5. Install an emulsion-base tar or drive adhesive over the entire driveway surface; cover everything.
6. Check the drainage along the sides so that the edges are protected.
7. Install 2-1/2 inches of a virgin mix of asphalt over the entire driveway. Check it yourself. Also look at your contractor's receipts to make sure the proper amount of asphalt has been installed. One ton of asphalt, 2 inches thick, covers 80 square feet. Do your math. Take whatever number of square feet you have, divide by 80 and you will get the tons he should have installed. Make sure the receipt is dated for your date of installations with your asphalt. If you still feel uneasy, call the asphalt plant to make sure he purchased from them, and make sure the plant says it is virgin asphalt.
8. Compress with a power roller and compact the edges with an iron tamp. Use a hand roller only for the first five feet, after that use it only to take out hot lines left by the power roller.
9. Check your driveway for drainage to make sure water can run off. You may see a few bird baths;

NOTES

put a quarter down, if the water covers the quarter, you have a puddle, not a bird bath. If you do find puddling areas in your drive, have the contractor patch it neatly so you can barely see the patch. This will save you money by ensuring that you won't have so many repairs down the road.

I enjoy patching. I take pride in my finish, so should the next raker and contractor. You won't get cheated if you follow these steps. They will save you money on repairs and you will be happier down the road.

You now have the same or more knowledge than the average contractor. Don't cheat in installing your driveway by going too fast. It is one of the most expensive repairs you will make on your home. Take all the time you need and get a good contractor--one who does physical work along with his guys is usually the best one to get. Study these steps, then compare them to your contractor's notes and see how you come out.

People have complained about my work at times. It makes me wake up, get smarter, and adjust. I still get a few complaints every now and then; I don't like them, but every day I find ways to help driveways hold up longer. If you do quality work and eliminate the mistakes, the complaints will disappear.

7

REPAIRS & MAINTENANCE OF AN ASPHALT DRIVEWAY

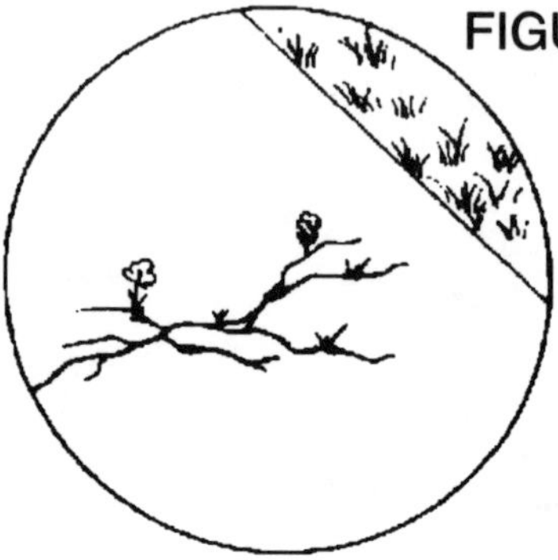

FIGURE 36

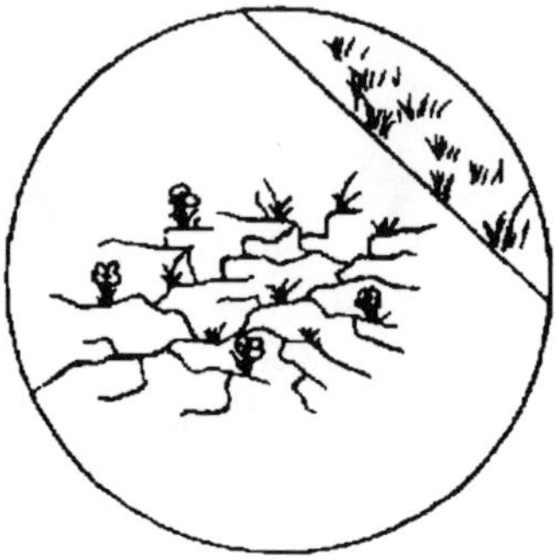

FIGURE 37

FIGURE 38

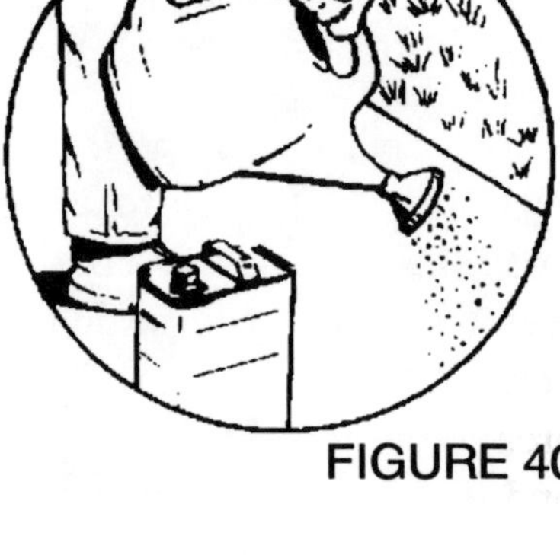

FIGURE 40

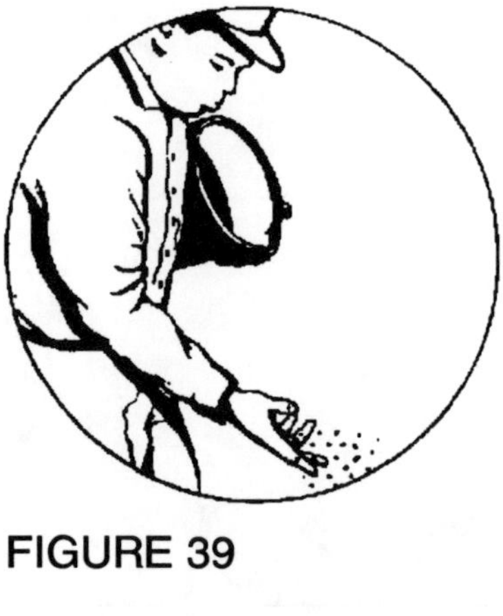

FIGURE 39

7

REPAIRS & MAINTENANCE OF AN ASPHALT DRIVEWAY

Small Repairs and Preventive Maintenance—Tools and Materials

If you have an older driveway, you can do a variety of things to keep it up to par. In some communities home owners may receive a notice from the building inspectors to repair their property. Often the driveway is part of the required repair. Now you are faced with the possibility of spending a good chunk of money. Let's say you have an asphalt driveway that is in bad shape for any number of reasons.

Reducing Grass and Weed Damage

For example, grass has overgrown the drive and there are cracks under the grass. Once you remove the grass and clean the driveway, you already have made a 24-40% improvement. I recommend that you use a good weed and vegetation killer on the area. If weeds and grass find their way through once, they will do it again with no problem. My father used to use table or rock salt along with the weed killer.

Once again I am talking about GOOD PREPARATION before you start spending money on products that will cost even more money than the weedkiller and salt, sealcoat and cold patch. If you have a contractor coming in to install hot asphalt in the patch area, you can save money by removing and killing the grass and weeds and filling in the cracks yourself.

Going back to removing the grass and weeds, you can use a sharp pick to get down into the cracks to lift out the weed roots. The pick will get down to the very root of

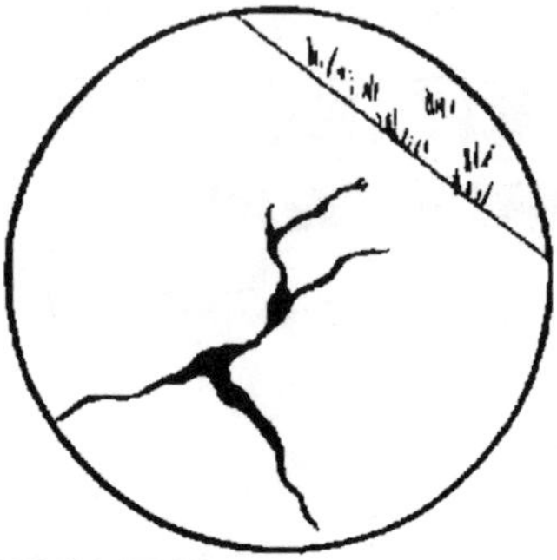

FIGURE 41

FIGURE 42

FIGURE 43

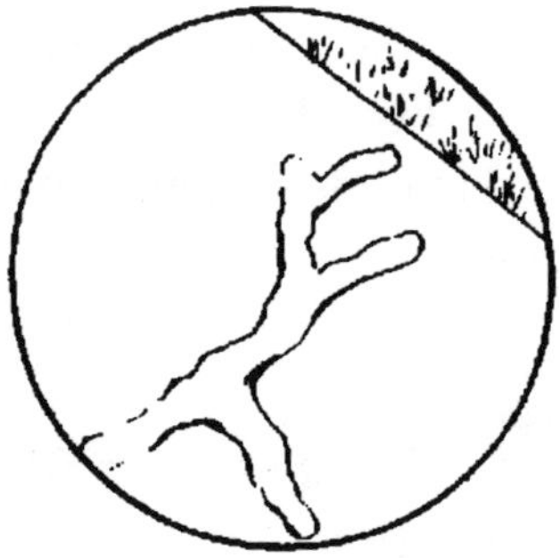

FIGURE 44

FIGURE 45

FIGURE 46

the problem (*and a problem is what you have*). You can pour salt and weed killer on an area time after time and they will not solve the problem if the roots remain. (See figures 36-40 and SBS Guide, "Weed Removal.")

Once you have the cracks cleaned out and apply the weed killer, you can move on to using a good crack filler in all of the cracks. Fill the cracks to the very top, then use a small flat shovel to distribute the crack filler evenly throughout the entire crack. Use the back of the shovel to flatten the crack filler even with the top of each crack. (See figures 41-44 and SBS Guide, "Asphalt Cobweb Cracks.") If it is applied too thickly, the crack filler will create other problems: You will track it in and out of your house; you may stumble over it; and if a snowplow scrapes the top of the crack, the entire area may be lifted by the blade. It's important to make sure you fill all of the cracks that the crack filler can get into. Open cracks will spread like cancer and will only get worse if you don't catch them in time. Water seeping through the open cracks will begin to damage the base-then you need a contractor for an even bigger job and that means more of your money. *That's what we are trying to avoid here. Protect your surface and you will protect your bank account.*

Use a cold crack filler with some type of rubber in it. The label might say "rubberized crack filler." It will last for years and will save money for you. While applying the crack filler, do it very slowly and easy. Just a little at a time will allow you to see how much more you are going to need. Use the back of your shovel to spread it as you go. By working slowly and smoothing as you go, the filler will also get deeper into the crack and at the same time you keep the area neat and save time (and time is money).

Patching Small Areas

If you have holes in your driveway that are not out of control, you can repair them yourself (see figures 45 and

FIGURE 47

46 and SBS Guide, "Emulsion, Tar Painting, and Sealcoating.") At your local hardware or building supply store you will find a material called "blacktop cold patch"; it comes in a large bag. Clean out the hole and use the same shovel that you used for the crack filler. (If any material is left over, save it for the next time cracks develop.) Apply the cold patch by packing it in a little at a time; use the shovel as though you were chopping meat on a butcher block. Do a little at a time and pack it as you go. Once you get to the top of the surface you are working on, you can use the back of the shovel to pound the cold patch down. One other suggestion, buy a $10-$15 tamp with a steel base at your hardware store. As you add cold patch, tamp it; then add more until you are level with the existing grade of the driveway. Would you believe you have spent less than a $100 so far, including the cost of the sealcoat, depending on the size of your drive (half what you would be charged for just one day of work by one person)? A contractor will not give you this information. He would convince you that you need a resurfaced driveway or hot patch and crack filler and sealcoat. A contractor would charge you $300-$500 instead of less than the $100 it costs you to do a better job yourself.

Sealcoating

Now that you have completed the basic repair work on the driveway, you can move on to dressing. Sealcoat is used to cover all of the work that's been completed so far. Before you paint new or resurfaced wood on a house, you have to prime it to protect it, or you know what Mother Nature will do to it! The same applies to the driveway surface. It is exposed to the outdoors and heat and cold will dry it out. Once it becomes brittle, it will start cracking. Sealcoating gives the driveway a nice film-like protecting coat. Like priming on wood, this protects the surface from the weather and keep it looking good for years. (See figure 47 and SBS

NOTES

Guide, "Emulsion, Tar Painting, and Sealcoating.")

One trick I have found useful is: while mixing a 5-gallon pail of sealcoat, you can pour 4-6 ounces of silica sand in it to form a good solid base for the sealcoat before you apply it to the driveway. The sand will also help to close the pores in the asphalt (it forms a slurry-like effect).

Always keep in mind, while preparing the driveway for sealcoating, that the area needs to be **extremely clean**! Scrap off anything that is not supposed to be there, gum, mud, all grass, clean the oil spots, etc. Fill the cracks, even along the edges of the driveway. Once you have done all this and the grass has been trimmed away from the edges, you can begin your final preparations for sealcoating. The first thing I do is to take a hose with a regular nozzle sprayer and spray the driveway **thoroughly**. Then use a good pushbroom to sweep off all the remaining debris. Doublecheck that the driveway has been cleansed thoroughly. Now check the square footage of the driveway so that you have just the right amount of sealcoat. You need about one gallon for every 50-100 square feet, depending on how you mix it.

When you take the lid off the sealcoat pail, you will see some water at the top; pour 4-6 ounces of silica sand into the 5 gallon pail and mix VERY SLOWLY, until you see a good even mix of product. Now, before you mix another 5 gallons, spread the first bucketful and let it dry about an hour in the sun to see whether you like the effect. Try mixing and spreading one gallon without the silica sand in it to see whether you like the look better without the sand (some people do). Since I'm more concerned with the longevity of your driveway than with the look I always recommend using the sand, but you may feel that the appearance is more important.

To apply the sealcoat neatly, you can use a paint brush to paint the edges as well as the first ten inches from the garage or the street. Use either a squeegee or a fine-bristle pushbroom, depending on the finish that you are

NOTES

looking for. In the city most people prefer using a squeegee; in rural areas most people are not so concerned with the appearance of their driveways. It's your choice, but remember preparation is the most important part of repairing your driveway.

A driveway is going to crack regardless of what you do to it. Two or three hours spent sealcoating your driveway surface every year or two will keep maintenance down. The total cost of sealcoating adds up to less than $10-$12 per month depending on the size of your driveway (under 500 square feet, the cost will be about $5.00 per month). Keep in mind that you are spending this money to maintain a quality driveway and that it is a lot less expensive than installing an entire new drive.

Some customers like the blacktop look of a driveway so much that they sealcoat too often, which can also hurt the driveway! They get a build-up and in cold weather, the drive becomes slippery. Too much sealcoating can also cause hair-line cracking. So be careful in monitoring your driveway and sealcoat only as you need it. Watch and see how dry and brittle it becomes. Watch and see how the first coat of sealer took; if you don't like it, find the brand that gives the look you want and that satisfies you as a consumer. *It's your money.* Don't spend over $7 per 5-gallon pail for sealcoat; try and get it on sale. Buy it from a good outlet store that you can trust.

Using a Contractor to Sealcoat

If you do decide to have a contractor install your sealcoat, then you now know what to look for. Make sure you watch how he mixes the product. *No more than 5 ounces of water to a 5-gallon pail of sealcoat, period!* I don't care what he or she tells you. Don't water down a product and run your money down the drain (or in this case, the sewer). Make sure he is following the correct steps of preparation that I spoke about earlier. Don't turn your back, and don't let the contractor

NOTES

sealcoat if the sealant was mixed before it arrived at your home. Have it done in your presence. I recommend that you use two coats of sealcoat the first time around. Once again, you are priming the drive and the first coat will soak right into the asphalt. The second coat will require only half the amount used for the first coat. Check the weather and make sure there is no chance of rain overnight, especially if you have some shaded areas where the sun cannot reach sections of the driveway. The second coat can be applied as soon as the first coat is down.

Paint the edges with a paint brush after you have filled the cracks. The edges crack first. The oil in the blacktop will break down over a period of years and the asphalt will lose its strength at the edges as well as on the surface. So, it's important to protect the entire drive as much as possible.

A driveway that has never been sealcoated becomes grey instead of black. That's because the oil has been evaporated by the sun and the blacktop begins to break down and become weaker. Sealing will help protect against some light oil leaks, but not a heavy downpour of oil. *Sealing will not protect against gasoline leaks!* Don't let a contractor tell you differently.

Protecting Against Gasoline Leaks

Let's talk about gas leaks for just a few moments. Here is what I suggest you can do to keep the driveway up until repairs are made. If the gas leak is not too large, you can repair the damage yourself. Cut out the area in a square; you can buy a masonry blade for your saw at your local hardware store. After you have completed the cutting, take a pick and dig down as close to the bottom as you can get. Get all of the gas-soaked asphalt and earth out, because it will weaken the area around it. Once you have done that, find a contractor in your area, take a pail, and buy a few shovelsful of asphalt. Go home, fill your area to the top,

NOTES

and take the flat tamp we spoke of earlier and tamp a little at a time until you have it **packed solid**. Let the area set a few days. After a few months you can sealcoat this area also. If a gas leak happens during the winter and you can't find hot asphalt, then buy the cold patch that comes in a bag and do the same thing. Once the weather is warmer and the asphalt plants open, fill the area with hot asphalt.

This takes me back to the beginning of this chapter. Always try to use hot asphalt in your holes even if they are not too large. It costs the same as cold patch and is *compatible* with the material that is already down. Hot asphalt will hold up much better, although cold patch will save you time and it will hold up for some time.

8

CONCRETE DRIVEWAYS

NOTES

8
CONCRETE DRIVEWAYS

Materials and Methods for Building a Concrete Driveway

Concrete: A mixture of cement, sand, gravel, and water that hardens as it dries. It is used for foundations, whole buildings, sidewalks, roads, dams, and bridges.

Cement: A fine gray powder made by burning a mixture of clay and limestone; Cement is in essence, a mixture of the silicates and aluminates of calcium, and is one of the most important of the modern-day silicon-bearing products. This substance, mixed with water, sand and gravel or crushed stone to form mortar, is used to hold stones and brick together in the walls of buildings.
--Definitions from *World Book Dictionary*
(Doubleday & Co., 1989)

Concrete is a solid. Cement is something you mix with other materials; it is a powder and does not become solid until it is mixed. Even though I work mostly with asphalt, I know from experience that concrete is the better of the two products. It is also the more expensive.

A concrete driveway will cost twice as much to install as an asphalt driveway, but it, if done properly, will hold up better and need less maintenance.

I first worked with concrete in 1970. My father was pouring his basement floor. The first thing I noticed was that concrete was prepared differently from blacktop. You had to have some knowledge of carpentry to set up the forms. Forms are a must when you pour concrete and how you set your forms is very important. Concrete is very heavy,

NOTES

so you must use sturdy supports for the forms. I noticed my father setting up first, even before he graded the area for pouring. You have to make sure that the area can support the stakes you will use in front and back of the forms. Once the concrete rests against the forms and begins to set, its weight will push against the forms. You can see the forms have to be supported very well with stakes or you can use metal forms that you can rent. The ground has to be solid to support the concrete. This brings us to the subject of installation of concrete driveways.

Preparing the Foundation

When preparing areas for concrete, you have to keep in mind your base. It is the key to a long life for your driveway. The area under the base must be extremely dry, NO MUD whatsoever in the area that you are going to be working. Once the area has been graded and all grass and weeds have been removed, you can install the drain if you are going to have one. Make sure that it empties into the street or is connected to a downspout that empties into a drain in the house. Be sure that you are not trying to run too much water to one drain. If you have had a problem with the house drains in the past, then don't tie into the house, go into the street. A 12 or 14 inch top with a 6-8 inch bottom (the neck of the drain or crock) is adequate for a 1,000-1,500 square foot driveway. You can usually have a 4-inch opening coming off the side of the neck from which you can use PVC piping (hard plastic) to tie into the street drain or downspout on the side of the house. Make sure that you cement the opening where you tie in the PVC into the clay drain. Buy a *fit strainer* for the drain to keep debris from falling in.

Just as for an asphalt drive, it is important that the edges of a concrete driveway be kept dry. Lay a 4 inch PVC pipe with holes in it along both edges. For an average 100-foot drive, you might spend $150; you can do the work

NOTES

yourself. Dig a trench, 6-8 inches deep, in the ground beside the drive; line it with small pebble gravel as a bed for the pipe. Once the drain and PVC pipe are in place, you can install the next important thing: the stone base.

When I interviewed a cement contractor, I asked him what he thought was a good base for concrete? To my surprise, he said that a properly laid base that had been previously covered with asphalt was the best he ever found. He went on to say that whenever he removed the asphalt the base underneath was always solid. He had my undivided attention! I wanted to find out more because I have noticed that a lot of cement contractors use only a little stone under the cement and I used to ask, "why?" They felt since cement was a stronger product than asphalt, they did not need as much base as asphalt required. That is not true, you need a good solid base, period. That's why the cement contractor I interviewed said a base that had been covered by asphalt made the best base for his concrete. He also mentioned that it does not crack as fast and that the drives he went back to look at never had to be repaired. I found that to be very important, you see. If you have a solid base for your concrete driveway and make sure it is packed solid with a driveway roller or a power tamper, you will save money and spend less time on repairs. **I hate to see money thrown away. Anyone hates to spend money on a major project in or around the home only to find in a short time the same job needs to be redone.** That's one of the things that got my attention and made me realize that this book needed to be written.

I once installed a friend's driveway and it didn't hold up the way I wanted it to. I returned and installed the driveway again. Well, it still did not hold up as well as it could have. I looked back to see what I had done wrong. I traced the problem back to the base. I had gotten away from what my father emphasized-*the material underneath is what supports the material on top!* Don't shortchange your investment or let a contractor do it to you either. *Does a*

FIGURE 48

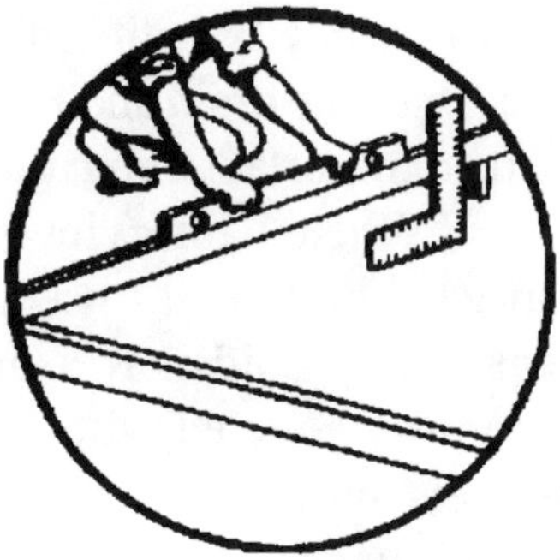
FIGURE 49

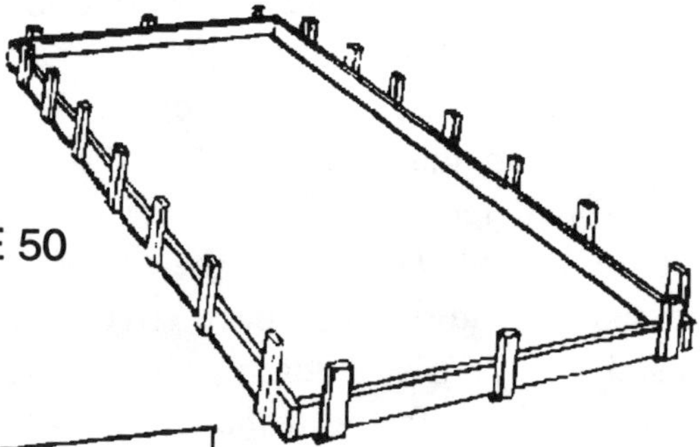
FIGURE 50

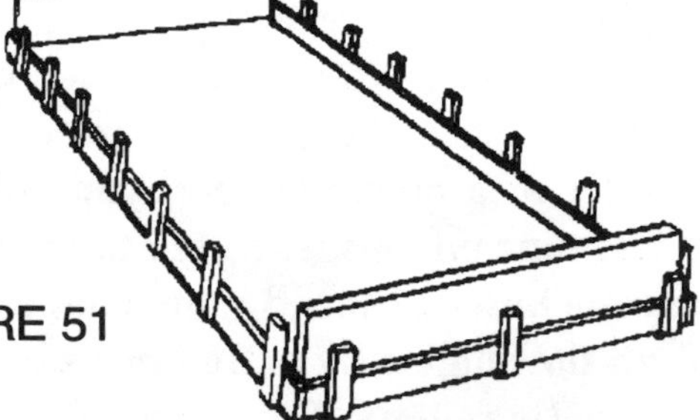
FIGURE 51

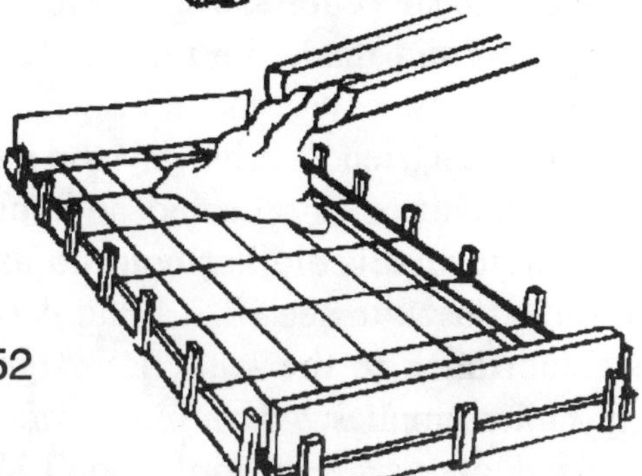
FIGURE 52

house stand without a good foundation? The same principle has been true since man first started to build. Concrete is no exception to the rule. Don't let a contractor install your driveway on top of 4 inches of slag unless you have at least a 12 inch sub-base under it. I recommend at least 6 inches of compacted stone.

At my local library I found some articles under "Highway and Heavy Construction" and they brought out some very interesting points on the sub-base and the top base. They said that you have to have a thick, solid base to start with. If not, what do you have to compress? Nothing! What do you have to support the heavy concrete? Nothing! So, once again, before moving on, don't get cheated out of your base or you will be cheating yourself out of money. The cement contractor I talked to also mentioned that if you pour too close to the ground and water gets underneath, when the frost comes and the ground freezes, the concrete will crack very fast. *And your money goes down the drain with it!*

Forms

Placing the forms is the next step. Use a level to grade the forms to allow water to run the way you want it to go. Proper setting of your forms is very important in forcing run-off. YOU DON'T WANT WATER SITTING ON YOUR DRIVEWAY.

Having the correct amount of forms is also very important. You can buy wood forms from your local hardware, building supply outlet, or a do-it-yourself store. Set them every 8 feet; if you don't have enough, it is okay to go every 6 feet. Keep a level on the forms to ensure that water will run off and away from the house. Make sure the water runs into the drain or off the side, and especially make sure it runs off the driveway. The forms hold the concrete in place and allow you to level it so that you don't get a wavy surface.

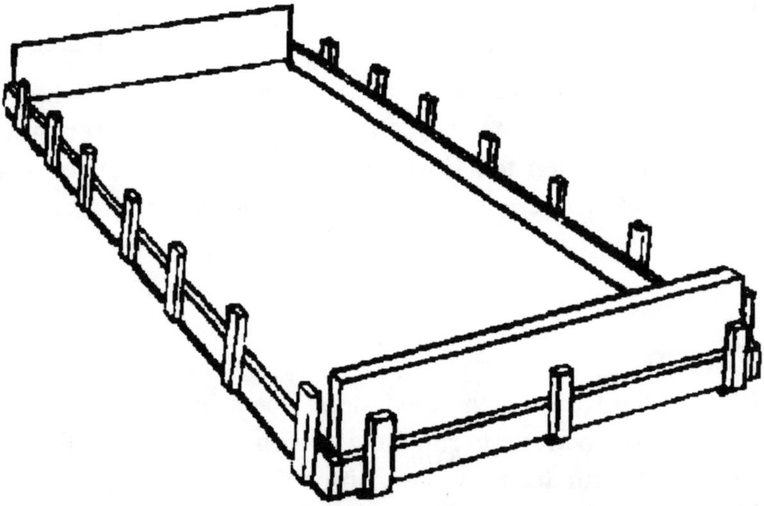

FIGURE 51

Now, a driveway needs to be 4 inches thick and the apron needs to be 6 inches or more. The blocks are poured to a depth of 6 inches right above the apron in my area. Remember that 2 X 4s are not a true 4 inches wide, be sure to buy wood that is a true 4 inches. Some municipalities have changed their codes to say they want 4-1/2 inches on the driveway because most contractors use the ordinary 2 X 4s. (See figures 48-52 and SBS Guide, "Concrete Form Building.") Once the the base has been installed and the forms have been placed, call your city inspector for a complete inspection before you pour the cement. But, not so fast! We have one more thing and it is very important. You must lay **wire mesh** into your driveway forms before you begin to pour the concrete. It will help keep the concrete in place. NEVER pour a driveway without wire mesh! It will not hold up (see figure 52). If you are not required to have an inspector's approval before pouring, then YOU make sure YOU inspect the driveway before it is poured!

Take a ruler and a 2 X 4, rest the 2 X 4 across the driveway forms (see figure 49). Then take your ruler and make sure that you measure the middle of the driveway as well as the sides to see that you have a true 4 inch depth across the driveway. Measure from the ground to the bottom of the 2 X 4 to get your true measurement. This is another reason that the slag or gravel base needs to be flattened or rolled to a proper compression-so you can get an accurate measurement. Install an expansion joint across the front of the form to separate the new concrete from the old concrete or asphalt and to allow room to expand. Then you should have an expansion joint every twenty feet or so (see figure 51). If your driveway butts up against your house, make sure you have an expansion joint along the house as well. All of these steps are important whether you or a contractor is installing your driveway. Once you have your driveway forms properly set up, you can call the local cement workers union and ask for a concrete finisher to

NOTES

come out and pour the cement for you. You will save a lot of money by doing it this way.

The Mix

The standard mix of cement for a driveway is called a "6 sack mix" of concrete. This is what you ask for when you are ordering the material. It is made up of 6 bags of cement plus gravel, sand, and water to make one cubic yard of concrete. If you want a stronger mixture, you can order a "6 and a half mix" or a "7 sack mix." Bridges and some streets and expressways get a different mix of material. If your contractor is doing the ordering, check to make sure he is buying the best product for you. Check your drive's square footage, then call a cement manufacturer or plant and explain that you want to know the quality and amount of material needed. If the company is good, someone will tell you how much and what grade you need. They want to keep your business.

Most cement plants know where you can find a good cement finisher or a quality contractor, if you go that route. Please keep this in mind: **You now know what to look for if you decide to go with a contractor. He can't tell you something that is not true.** ***You will know what to do or look for yourself! The end result will be TO SAVE MONEY ON YOUR ASPHALT OR CONCRETE DRIVEWAY.***

9

MAINTENANCE OF A CONCRETE DRIVEWAY

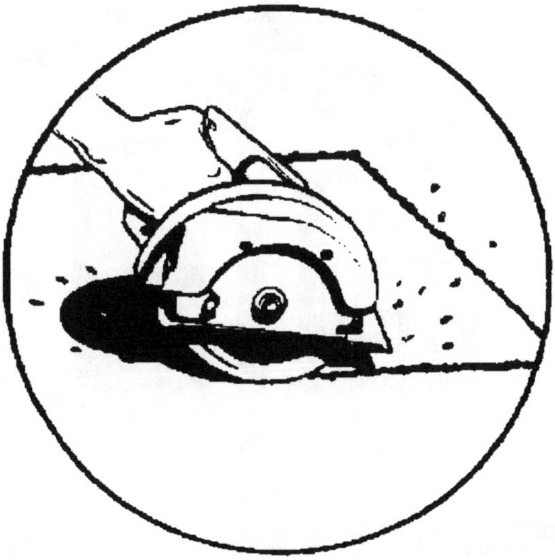

FIGURE 52

9

MAINTENANCE OF A CONCRETE DRIVEWAY

What to Do to Make a Concrete Driveway Last

The worst enemies of a concrete driveway are SALT and FROST. People who specialize in installing concrete driveways are in total agreement. Salt causes concrete to "pit." So don't put salt on your driveway! Aside from salt problems, a concrete driveway is maintenance free. You have to watch out for cracks. If one does develop, you can fill it with tar. If you have a section that falls apart, a contractor or you yourself can cut it out using a masonry blade available at most hardware stores (see figure 52).

Now let's say it "pits" all over because you got a bad mix of material, then you can always use the driveway **as a base in an installation of a new asphalt driveway.** That's what is done on the highways. You can blacktop over concrete, but you cannot pour concrete over blacktop. If you decide to go with asphalt, make sure the cement is solid and the area has been prepared. If there are cracks, you can have the blacktop contractor install a "petromat fabric", a nonwoven fabric that looks like gray carpet (the asphalt plants carry it). It is used to cover the cracks and will help slow the cracks from coming through the blacktop.

The steps that I have mentioned for resurfacing an asphalt driveway apply when you cover a concrete driveway with asphalt. Of course, now you have concrete to serve as a base instead of asphalt. An asphalt covering will be less expensive. Keep the new asphalt surface in good repair and seal all cracks with hot or cold-base tar and you will be okay. Or you can, if you want, pour another concrete driveway.

If your concrete driveway does become badly pitted and begins to fall apart and you have been careful not to use salt on it, go to the concrete plant first and complain. Don't cover it with asphalt until you see what they will do for you. They may refund your money.

10

CHOOSING A CONTRACTOR

NOTES

10

CHOOSING A CONTRACTOR

Finding an Honest Contractor and Keeping Him Honest

One of the key concerns to check in selecting a contractor is whether the contractor is registered in one or more municipalities for at least 3 years, and has been installing asphalt or concrete on his own for at least 5 years. He needs at least that much experience to make sure he has a good working crew. His crew needs a chance to work together and know what they are doing as a team; their work will speak for itself. *You want a good crew with a sense of unity on the job when they work around your home.* Often the contractor will leave his workers alone for a period of time and they have to have unity and be able to work together. That might not seem important, but it is. They need the work experience as a team.

Another thing to watch for is whether the salesman comes out on the day of the job to go over the things you and he agreed upon. I have had three salesmen at one time and know that they need to work with the guy running the crew to ensure that you get what was promised in your contract. Insist upon it. Often workers will know more than the salesman and will want to do the job different from the way the salesman explained it to you. Everyone has a different theory on how the job should be done. You, the salesman, and the workers may all have different opinions about the boundary, the width, and where the driveway should start. Having all of the parties there at the same time will cut down on mistakes.

Be sure to check the contractor's insurance and bond policies to see that they are valid and in force. Tell

NOTES

the contractor that you have to see the building permit before any work starts. And tell him that you want all receipts to have your address on them and to be marked with the date of delivery for your job and that it says paid in full. You don't want the suppliers putting a lien on your property if the contractor should go out of business.

Another thing you should do is to double-check any references the contractor might give you. GO OUT AND LOOK AT HIS OR HER WORK! See if you will be happy with their finished product. Find out when the work was installed and how it is holding up. Sometimes a drive, whether it is asphalt or concrete, will start to develop hairline cracks within the first through the fifth year; don't let that alarm you. What you don't want to see are large openings or heavy cobweb cracks, which often mean that the problem is below the surface and that someone did not do his job. Maybe the contractor used all recycled asphalt, which I have found will not hold up the same as virgin asphalt. Some recycled asphalt mixed in with the binder is okay, but it is not suitable for the top layer. Check to see if the sides have been tamped solid and neatly. See whether grass has overgrown the edge and ask if it's okay to pull a section back (remember the edges are important) to see how the edge is holding up. Ask earlier customers whether the contractor was willing to make an appointment to discuss any problems. Were problems caused by the contractor repaired in a reasonable amount of time? How did his crew act while working around the house? Ask the previous customer all of these questions.

Check out his equipment; ask to see pictures. Check the material receipts to see that they are marked paid for your job. Ask that your address be on all receipts before you sign the contract, so the plant cannot put a lien on you later. Ask for a copy of his business insurance, don't accept his word that he has insurance. Make sure the contract is clear and explains everything in writing. Get a written promise that he will be back in 30 days or less if there is a

NOTES

problem.

These are things I had to learn through my own mistakes. I still find it hard to get good help and it is hard on me to follow all of these guidelines. But as a contractor I must make these sacrifices and be hard on myself. Customers made me and are making me better. That's why I resolved to be a repair contractor. Now I know I can find two good guys and do a better job than when I had to work with 5 or 6 inexperienced people. I can control the sealcoating and asphalt repair. I hope to continue to offer a good repair and sealcoat job and to install asphalt and concrete driveways that both my customers and I can be proud of.

11
NEW MATERIALS BEING TESTED

NOTES

11
NEW MATERIALS BEING TESTED

New Materials to Improve Asphalt and Concrete Durability

One thing that helps asphalt last on resurfaced roads, highways, driveways, or parking lots is to lay a "petromat fabric" between the old surface and the new finishing layer. This covering, which looks like a gray carpet, helps retard cracks from coming through to the surface. The theory is that this nonwoven fabric, which is a asphalt polymers formed by a strong chain of molecules linked by their carbon atoms, provides the flexibility needed for asphalt to endure both freezing temperatures and the heat of a summer day. The polymer develops a lattice in the bitumen.

> To prevent cold weather cracking, ciastomeric polymers such as natural latex styrene butadiene rubber, styrene/butadiene block copolymers and neoprene can lower the temperature at which the asphalt will break from 32°F to 10°F, and that makes the asphalt more rigid and stronger and will help prevent cold weather cracking. The goal is to extend the asphalt pavement's performance properties throughout the anticipated working temperature, which range as high as 140°F to as low as 20°F and they (the polymers) are expected to remain pliable at processing temperature and adhere to the aggregate better.--Quoted from *Highway & Heavy Construction Magazine*, December 1989.

Some states, such as Colorado, specify polymer

NOTES

modification for nearly all road repairs. The Federal Aviation Administration's runway specification is the only one that mandates polymer modification (source, *Highway and Heavy Construction Magazine*, 1989).

A Piece on Concrete

Howard J. Durham and Will K. Secre said, in *Engineering & Technology*, that today's concrete mixtures offer a new freedom in project design, concrete delivery, and construction. They contain water reducers, retarders, accelerators, superplasticizers, and mineral admixtures. These are new products now available in all traditional concrete mixture classifications. In addition to traditional benefits, the new products provide good finishing ability and give better hydration control. The water reducer helps the concrete to be placed and finished faster, saving both labor and time. Contractors should pass the savings provided by this new technology on to their customers. In my opinion, everyone should benefit from cost reductions.

CONCLUSION

CONCLUSION

I want to thank my father for the experience of watching him work and teaching me what he knew. The extent of his knowledge and the quality of his work meant that he never really needed to advertise for work. What I saw him do I have seen work for me over the years. My dad worked on over 15,000 driveways and taught my five brothers and me how to do the work right. He has done a GREAT job in what he accomplished. He and his partner installed more drives and parking lots from 1969-1974 in Chardon, Ohio, than anyone else; all without advertising. In 1979, dad went back home to Alabama and then to Houston, Texas. From time to time he returns to help my brothers and me and to give us some very good advice. Daddy, I did and still do listen to you!

In closing, take with you this advice: Don't be afraid to ask questions, to challenge the contractors, to have your specifications in writing, to question your costs--you are well within your rights to do so.

I hope that the material in this publication will help you and your household for many years to come. Follow the advice given here and you will save money. This book gives you enough insight to make the proper decisions and to understand what contractors tell you.

I have enjoyed sharing the professional and personal side of my years as a contractor. Thank you very much for your support!

--Al Edwards

STEP-BY-STEP GUIDE
QUICK REFERENCE SECTION

ASPHALT & CONCRETE SUB-BASE

Dig a trench 8-16 inches deep, down to solid ground, for the sub-base. All mud must be removed and both a weed and vegetation killer must be applied.

The sub-base stone should be 2-4 inches in diameter and be spread in a layer at least 6 inches thick.

If you are doing the work by hand, have the stone dumped in small piles along the length of the driveway, then level the piles with a shove.

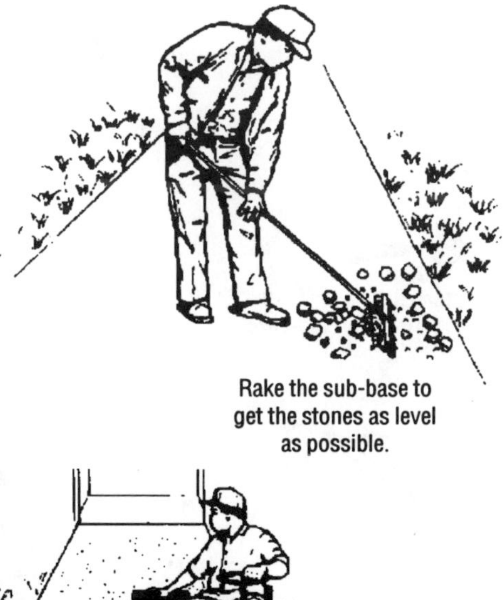

Rake the sub-base to get the stones as level as possible.

The sub-base must be compacted with a heavy roller before the finish base is installed.

ASPHALT & CONCRETE FINISH-BASE

If the base has been down for a while, check to be sure there are no soft or muddy spots. If there are, dig them out.

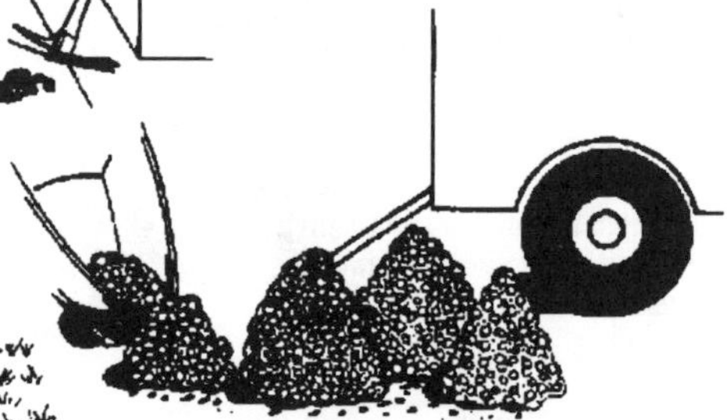

For the finish base, smelter limestone rocks, 2-3 inches in diameter, are perfect.

Spread the stones for the finish base evenly over the sub-base.

Rake the finish base to get the stones as level as possible.

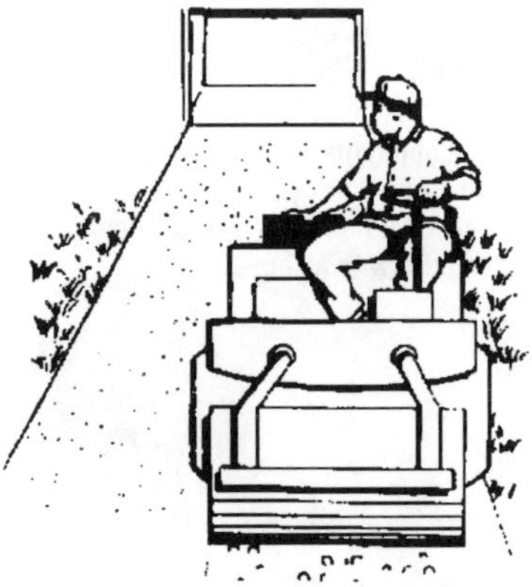

The finish base must be compacted with a heavy roller before the binder layer is installed.

4 inches of asphalt, two layers

6 inches finish base, crushed limestone

8 - 12 inches sub-base

Solid earth - NO MUD

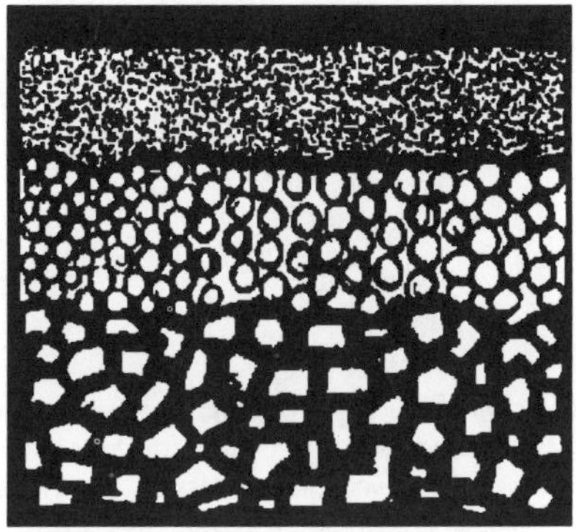

WEED REMOVAL

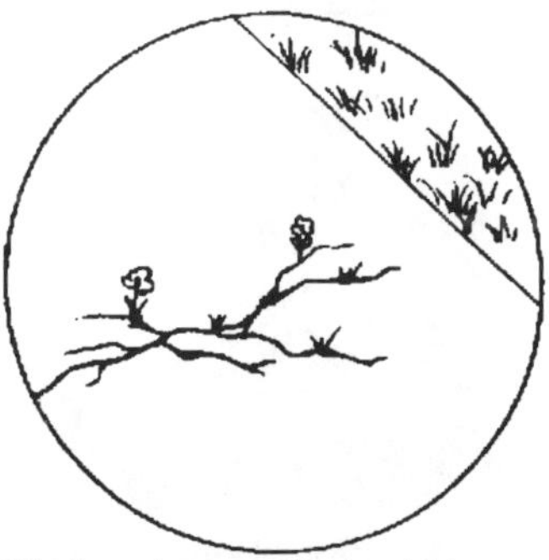

Weeds and grass will grow right through a blacktop surface.

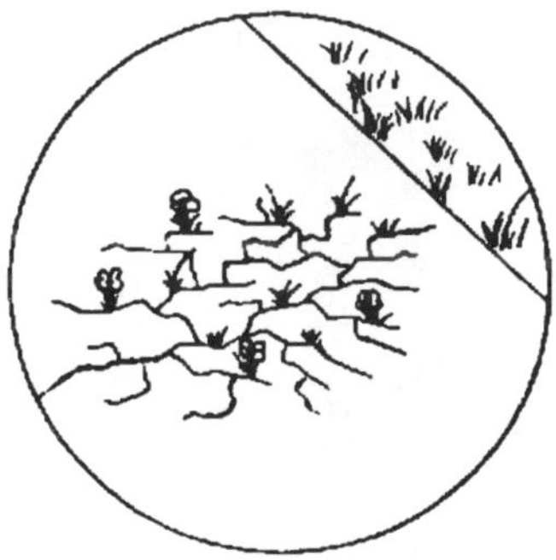

Dandelions are major problem if they are not removed carefully.

Use a pick to dig through the blacktop and down to the roots of any weeds.

Ordinary salt or vegetation and weed killer should be used to destroy grass and weeds. Salt is effective when covering a large area after mud has been cleared away. Liquid weed and vegetation killers will work if you have cracks in the asphalt.

ROLLERS

The power roller is the only roller that will compress the asphalt to a proper level.

The hand roller is used to take out hotlines left behind the power roller.

TAMPING

It is vital for the long life of your asphalt driveway that the asphalt be properly compacted. Along the edges where the heavy roller is not so effective, you or your contractor should use a heavy steel tamp to compress the asphalt. It will add to the life of the driveway and will also make it look neater.

CONCRETE DETERIORATION

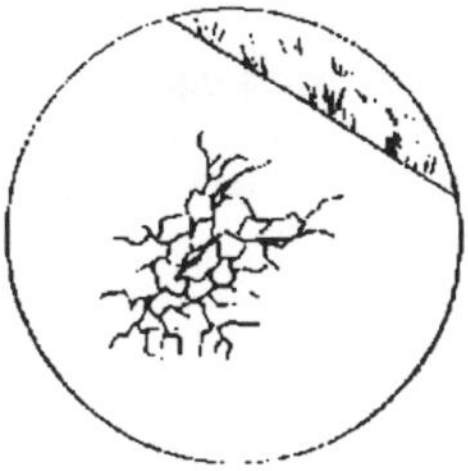

Cobweb cracks before repair.

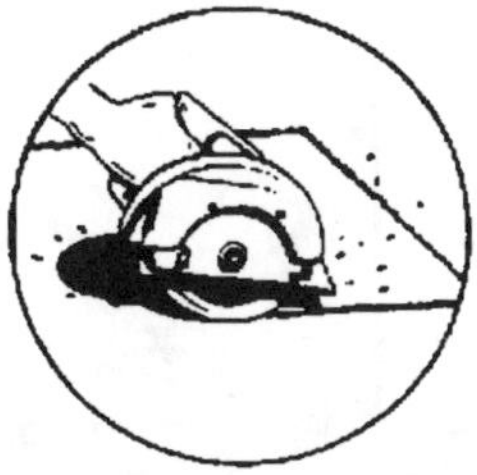

A masonry blade on a rotary saw can cut asphalt or concrete neatly.

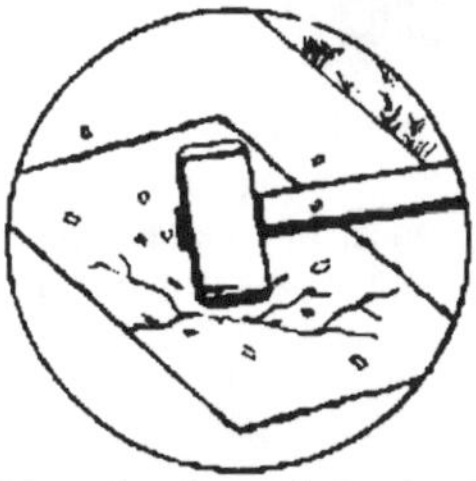

If the cobweb area is too hard or tightly packed to remove, a sledge hammer should break it enough to work with a pick.

Use a shovel to remove all loose rubble and mud.

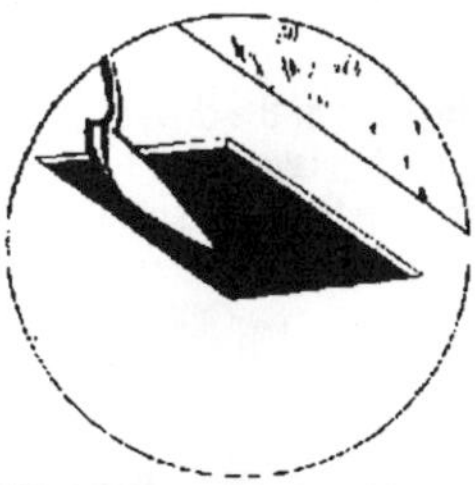

Fill the cleaned hole with hot or cold-patch asphalt.

Fill the area to the top, then tamp it level with the surface of the driveway.

ASPHALT COBWEB CRACKS

Use same procedure outlined above: eliminate masonry cut, sledge hammer break up, and shovel removal. These are not needed for asphalt.

All asphalt cobwed areas should be removed with a pick mallet.

LEVELING & REMOVING ASPHALT

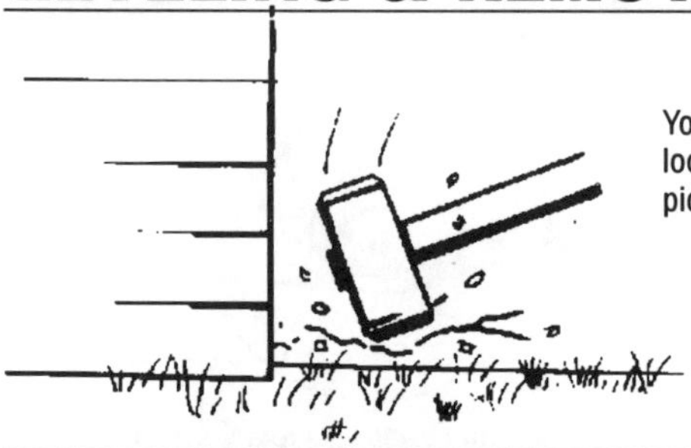

You can use your sledge hammer to loosen and level asphalt. Then use a pick and shovel to remove asphalt.

EMULSION & SEALCOATING

After all cracks and holes have been repaired, the driveway must be thoroughly cleaned, Mix the sealcoat. Spread it by using strokes that push it forward, then draw backwards to give a squeegee effect.

EDGING

You must trim the edge of the driveway before installing the top layer of asphalt. Unlike concrete, your asphalt is compressed an the side with a steel tamp.

EMULSION & SEALCOATING

Once you have dug out rubble and mud from an allegation crack area, use either hot or cold patch asphalt to fill the hole.

Use a flat tamp (one with a wooden handle and steel base is okay for this work) to compress the hot or cold patch asphalt and make it level with the surrounding surface.

CRACK FILLING

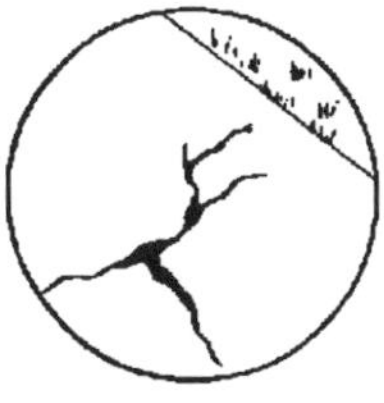

First, any bad cracks must be cleaned out. A pick mallet is a perfect tool for scraping out a crack and digging down to a drier layer.

After the cracks have been cleaned, fill them with crack filler.

Level and smooth away the excess crack filler by using the back of a shovel. This is necessary to prevent the filler from flaking or forming a rough edge to catch a snowplow blade.

The filled crack should be smooth and level with the surface of the driveway.

CONCRETE FORMWORK

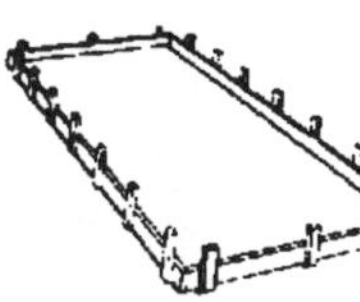

Drive the stakes at least 8"-10" deep. Nail from outside, not inside and through the stake into the 2 X 4 frame. Do not nail all the way in.

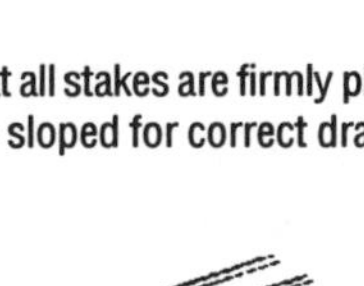

Use a level to ensure proper drainage toward a drain. Bottom form is at least 4 inches above the ground all across the driveway.

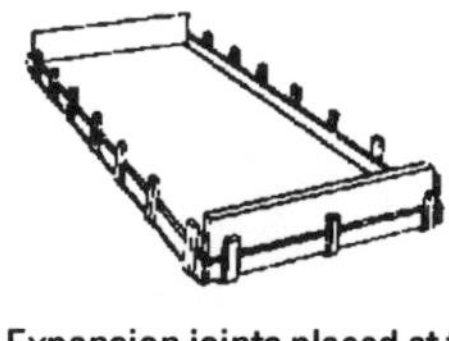

Double-checks that all stakes are firmly planted. Surface should be sloped for correct drainage.

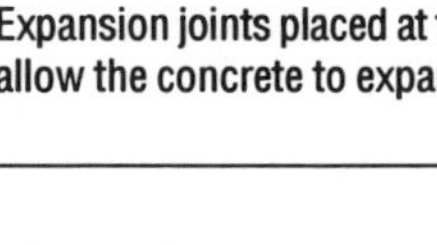

Expansion joints placed at the end of each form allow the concrete to expand in hot weather.

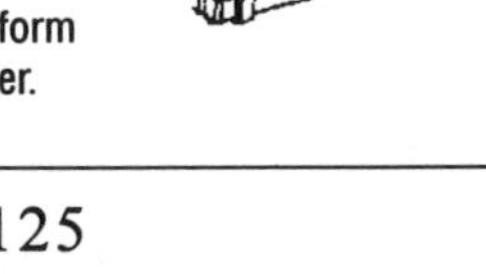

Do not allow a contractor to pour concrete without the wire mesh. This will hold the concrete in place.